SPACE EXPLORATION

Karen Hales Mecham

Advisors

Ilene L. Follman

Ann C. Edmonds

Primary Science Resource Guide
Includes Transparencies, Reproducibles, and Teacher's Guide

SPACE EXPLORATION

Illustrations: Donald O'Connor
Project Manager: E. Rohne Rudder
Cover Design: E. Rohne Rudder
Typesetting: Pagecrafters, Inc.
Managing Editor: Kathleen Hilmes

To The Teacher

I believe a leaf of grass is no less than the journey-work
of the stars.

Walt Whitman

For as long as people have walked the earth, the night sky has excited the human imagination, inspiring stories, poetry, and songs in every culture. The keen observations at the heart of these ancient stories served as the foundation for the later work of great astronomers and physicists whose scientific approach to observation and experimentation has enhanced our knowledge of the universe. We continue to explore the vast, relatively unknown wilderness of our universe through experimentation, observation, and exploration.

The information and activities in this *Space Exploration Resource Guide* are organized in roughly three sections: the **Space Travel Simulation; Our Solar System and Beyond; and Energy, Force, and Motion in Space.** Learning opportunities in each section are planned to engage children and teachers in experiences that allow for free exploration, concept development, and application of concepts. A classroom space shuttle simulation provides the focus for child exploration throughout the unit of study. Pretend space exploration stimulates curiosity, motivating children to research information about the solar system and investigate scientific principles at work in the universe. The activities in the resource guide are not organized in a sequential, lock-step way, but rather are structured so teachers can choose from activities as if they were selecting from a menu—planning learning opportunities based on children's interests and levels of understanding. The concepts of time, distance, and space are very abstract and difficult for many young children to comprehend. Therefore, activities outlined in this resource guide vary in general grade-level appropriateness allowing teachers to adapt them to meet their own students' needs. Opportunities to develop the science skills of observing, data gathering, classifying, measuring, recording, listening, and writing are numerous. Children can record observations, experiments, and investigations in their space journals. As children represent their understanding in words and pictures, they revisit a concept or experience to organize their thinking and construct their own meaning and logic.

A **To the Parent** reproducible page provides an overview of the Space Exploration unit and suggests ways parents can enhance their child's experiences. Four **reproducible pages** for recording data from research and investigations are included for student use. They are referenced as **R** within the text. A **Solar System Chart** provides teachers with a quick reference guide to information about the planets. This is meant for teacher use; children should be allowed to experience the excitement of discovering this information through their own independent research. Four **transparencies** illustrate concepts introduced in activities and can be used to engage students and teacher in discussion and exploration. They are referenced as **T** within the text. (See also inside covers.) The **bibliography** includes books and resources for children and teachers.

Start your countdown to a great learning experience in your classroom: 10, 9, 8, 7, 6, 5 . . .

Contents

© Milliken Publishing Company

To the Parent

In science class, we are beginning our study of the solar system and space exploration. In this unit, children will learn about the universe as they pretend to be astronauts traveling in space. Space shuttle simulation activities will allow children to "experience" working and living in weightlessness, eating and sleeping in space, and maintaining physical health during space missions. Children will research information about the solar system and investigate scientific principles at work in the universe. Investigating shadows, the sun, the moon, constellations, planets, night and day, and the seasons will enhance our understanding of Earth's place in space. We will study energy, force, and motion in space as we investigate gravity, energy from the sun, rockets, aerodynamics, and centripetal force.

Share in your child's exploration of space with some of the following activities.

• Talk with your child about discoveries made in class.

• Share books about space exploration, our solar system, myths*, and science fiction with your child.

• Take evening walks together, and observe the night sky.

• Visit a local planetarium, and observe the night sky through a powerful telescope.

• Call a local astronomy club, and join them for one of their "observe the night sky" outings.

• Participate in the parent-child homework activities.

• If your child shows a keen interest in space, consider the possibility of attending a parent-child weekend at NASA's Space Camp in Huntsville, Alabama or Titusville, Florida. This program is designed for children aged 7—11 and their parents. Call 1-800-637-7223 for information.

* Night sky myths to share with your child:
Book of Greek Myths, D'Aulaire
The Mending of the Sky and Other Chinese Myths, Xiao M. Li
Mythology and the Universe, Asimov
The Shining Star—Greek Legends of the Zodiac, Vautier
They Dance in the Night Sky—Native American Myths, Monroe and Williamson

Getting Started

Free Exploration

Build a child-sized space shuttle model in your classroom where children can simulate real space travel and scientific investigation. The dramatic play and exploration connected with the space shuttle model will stimulate curiosity and enhance student enthusiasm for the study of space.

Establish a display and exploration area in your classroom where children can research space topics of interest. Include books, photographs, models of the solar system, a globe, a moon map, color photographs of Earth taken from space, and so on.

Collect many of the books listed in the bibliography for use during your study of space. These outstanding books will provide children with many hours of pleasant reading and research.

Visit a local planetarium. Observing the night sky through a powerful telescope will enhance a child's fascination with space. For a listing of planetariums and observatories in the United States and Canada, see the *National Geographic Picture Atlas of Our Universe* by Roy A. Gallant, pages 280–281.

Read the Ancient Star Myths inspired in every culture by human fascination with the night sky. Children will enjoy listening to the ancient Egyptian story of the sun god, Ra, traveling daily in his boat from east to west across a heavenly ocean. Norse myths, Native American myths, and Hindu myths, to name just a few, will serve as an outstanding literature connection for the unit.

Watch NASA videos—*Milestones of Flight, Living in Space, Toys in Space,* or *Eating and Sleeping in Space*—to stimulate interest in space. See *Bibliography* for NASA address.

Order outstanding education materials and brochures from NASA. Most materials are free or available at a minimal cost. See bibliography for NASA address.

Connect your classroom computer with NASA Space Link, an electronic information system for educators at the Marshall Space Flight Center in Huntsville, Alabama, designed to communicate with computers and modems. Space Link is free, but you must pay for long distance telephone calls. Write to NASA headquarters in Washington, D.C. for information about Space Link. Students as well as teachers can access up-to-the-minute information about space through Space Link.

Keeping Records

Keep a space journal to provide children with a place to represent their science knowledge and understanding in drawings and words. The journal should become the child's companion throughout this unit of study—a place to record experiments and demonstrations, as well as student understanding of the concepts at work in these activities.

Reproducibles included in this book may be included in the space journal or kept in a separate portfolio.

Use the space journal for assessment. The space journal serves as an excellent assessment tool. Reading a child's description of an experiment or demonstration and her or his explanation of a concept provides the teacher with invaluable information about student understanding. Teachers may note special interests recorded in the journal and encourage a child to pursue an independent project. As teachers review student journals, they can learn about a child's misunderstandings and plan additional experiences to enhance learning.

Collect news articles about space. Display news on a bulletin board or in a book of space news.

Write a space newspaper—*The Daily Sun*—as a class, summarizing the news and information children collect as they conduct their independent research.

Technology and Space

Teacher Information

Telescopes... We learn about stars, planets, and galaxies by studying the light and other energy they radiate into space. To detect light energy, we can use our eyes, but optical telescopes greatly enhance our vision. Optical telescopes use mirrors, glass lenses, or both to collect, focus, and transmit light resulting in a magnified image of stars and planets. Infrared and radio telescopes collect infrared and radio waves emitted by objects in space. Because only visible, radio, and infrared energy waves can get through Earth's atmosphere, some telescopes are sent up on balloons, aircrafts, rockets, and spacecrafts to collect ultraviolet, X-, and gamma rays to help us learn more about objects in space.

Hubble Space Telescope... Released into orbit from the shuttle cargo bay, this telescope will orbit 500 km (310 miles) up above Earth's hazy atmosphere, allowing us to see seven times deeper into space than we can with any telescope on Earth. The telescope will be repaired and maintained by space shuttle crews.

Satellites...A moon or man-made body in orbit around a planet is called a satellite. Satellites in orbit around Earth collect and transmit information about weather and aid in pollution control, the search for oil, and the mapping of Earth's surface. Photographs collected by satellites give us important information about Earth. Communication satellites relay television programs and place long-distance telephone calls.

Apollo 11..."One small step for man, one giant step for mankind." The first person to set foot on the moon was Neil Armstrong. He took this historic step on July 20, 1969, during the flight of Apollo 11. Apollo 11 astronauts returned to Earth with the first moon rocks. Five more Apollo lunar landings followed, ending in 1972 with Apollo 17.

Space Shuttle... The space shuttle is the world's first reusable spacecraft. It is a winged orbiter that carries people and experiments to Earth's orbit. Astronauts travel in the shuttle and perform experiments in space. The shuttle may stay in space for a few days or weeks. Its velocity in orbit is about 28,000 km (17,500 miles) per hour. It circles Earth once every 90 minutes. The shuttle crew lives and works in the cabin. The crew controls the orbiter and handles most payloads from the top-level flight deck. The living area at mid-deck is the area where crew members sleep, eat, exercise, and take care of personal hygiene. The payload may carry communications, weather, and scientific or military intelligence satellites. A robot arm can be operated by astronauts to reach out from the orbiter. The robot arm can take a satellite from the payload bay and release it into an orbit around Earth. It can also reach out into space and capture a satellite in need of repair. When ready to return to Earth, crew members slow the orbiter so it can "drop" out of orbit and return to Earth.

Space Station...A space station is an orbiting structure that provides a working and living space for humans above Earth's atmosphere. Crews travel to the space station to work and then return to Earth when their work is done. A space station has many uses. It is an observatory where scientists can observe space without interference from Earth's hazy atmosphere. It is a biomedical laboratory for testing human response to weightlessness. It is a research center where we can test medicines and alloys at zero gravity. We might be able to manufacture products that can't be made on Earth. The space station can be used as a spaceport for launching spacecrafts on flights to planets and beyond. Orbiting Earth once every 90 minutes makes the space station a perfect place for observing it. Space stations launched in the 70s and 80s included the former USSR's Salyut 1 and Mir and USA's Skylab.

Other important space missions include: Sputnik, Mercury, Gemini, Apollo-Soyuz, Pioneer Venus 1 and 2, Pioneer 10 and 11, Viking 1 and 2, and Voyager 1 and 2.

Space Travel Simulation:
A Space Shuttle in Your Classroom

Build a space shuttle in your classroom where children can pretend to be astronauts and experience space exploration first-hand. This activity is designed to capitalize on children's natural curiosity about space exploration and their robust fantasy world. This pretend play enhances a child's enthusiasm for learning about space. There are many possibilities for building a space shuttle or spaceship model in the classroom.

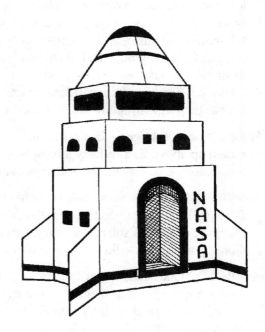

Stack three or four boxes of different sizes on top of each other. Cut a door in the biggest box and a hole in the top of it. Fasten a second, slightly smaller box over this box with the open side down. Make a cone from a circular piece of poster board. Cut a line along one radius, then curve the circle into a cone shape. Attach to the top of the space shuttle. Children can paint the space shuttle with white paint and draw a NASA insignia on the side.

A large refrigerator box with a cone shaped roof attached can be a spaceship. Cut windows in the side of the box and cover with clear plastic. Attach shuttle wings on the sides. Paint.

Convert the playground jungle gym or climber into a spaceship.

Create a control panel with various knobs, dials, etc. Children can design this panel and invent ways to use it during their pretend space travels.

Attach one side of a length of velcro along one wall of the spacecraft. Cut small pieces from the second strip length and glue the velcro pieces to items like pens, small notebooks, glasses, telescopes, silverware, razors, mirrors, toothbrushes, combs, etc. The velcro keeps objects from floating around in the spacecraft.

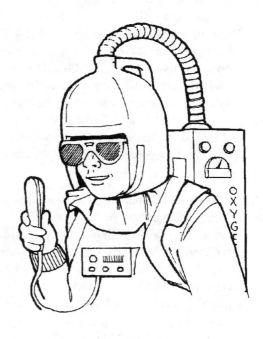

Design life support gear for "astronauts" to wear in extra-vehicular work. Make a space helmet and oxygen tank using plastic milk cartons. Attach a length of plastic hose from the helmet to a box with a shoulder strap attached for the oxygen tank. Empty plastic soda bottles can also be used for oxygen tanks. Children might want to use ropes for tethering themselves to the orbiter while working outside of the shuttle so they don't drift off into "space."

materials: plastic hose, plastic milk cartons, boxes, cotton rope

Play space music inside the classroom space shuttle, e.g., New Age recordings, *Star Wars* recordings, etc.

Space Exploration

Living in Space

Teacher Information

Food... People who travel in space must take food for survival. The food must be nutritious, lightweight, convenient to use, easily stored, and require no refrigeration. Some food is **dehydrated** to meet weight restrictions for liftoff. When astronauts are ready to eat something, they simply **rehydrate** it using water available on board. Water used for rehydration is available as a by-product from the shuttle's fuel cells which produce electricity by combining hydrogen and oxygen, resulting in water. Many foods like spaghetti, scrambled eggs, and strawberries go through the dehydration-rehydration process. A synthetic orange drink product is mixed with water and used in place of orange juice. Some foods like graham crackers, cookies, peanut butter, and gum are taken aboard in their natural form. Some foods are preserved by exposure to radiation. Meats like tuna fish are packaged in individual servings. Dried apricots, peaches, and pears are examples of foods from which part of the water has been removed. A liquid form of salt and pepper is used because crystals would float around the shuttle cabin. Silverware, trays, can openers, and cans are held to a table by magnets or velcro tape.

Health... Astronauts must stay in the best possible condition. In order to do this, they must exercise. Scientists discovered that astronauts on early missions suffered some bone and muscle deterioration because they were not getting the resistance they were accustomed to from gravity on Earth. Astronauts walk a treadmill for fifteen to thirty minutes a day to provide them with the resistance exercise they need. Astronauts must take special care in space to get proper food, exercise, and rest. They also pay close attention to cleanliness. Astronauts cannot shower in space because water floats around the cabin, so astronauts take sponge baths using washcloths to hold the water. The washcloths are stored in bags. A mission medical specialist provides astronauts with medical care. A medical kit is stored on board.

Working and Living in Weightlessness... Weightlessness controls how people live and work in space. The force of gravity on Earth attracts all things toward the center of the planet. This force gives objects their weight. In an orbiting spacecraft, the force caused by the forward motion of the spacecraft balances the pull of Earth's gravity, resulting in weightlessness. Objects in a weightless environment float. People float, computers float, food floats, water floats—everything that is not securely fastened, floats. Weightlessness also affects the body, causing blood from the lower part of the body to shift to the upper part. People are three to five centimeters (1-2 inches) taller because gravity is not pulling down on the vertebrae of the backbone. The knees bend and people walk in a slightly crouched position. People have to hold their arms down. Space boots have special attachments that raise the heel so the toes don't point up. Astronauts move around the spacecraft using hand-held suction cups. Suction cups on shoes, straps, and velcro strips also help astronauts move around and do their work in the weightless environment of space.

Communication... Communication is crucial for a successful space mission. Astronauts must communicate with ground control so information can be collected. They wear headsets with microphones. Astronauts must also be able to communicate with crew members in other parts of the spacecraft. The language used must be clear and concise. All data processing for space missions is handled by computers on board. The computers are programmed to control the basic running of the spacecraft. They process medical data from the crew, monitor the air content of the spacecraft, and keep the spacecraft on course. Every phase of the space mission is controlled by the spacecraft's computers.

Living in Space: Investigations

Investigating Working and Living in Space

Take a make-believe trip into outer space. While playing "space music" in the background, tell the children a story about space travel. Listen to the guided imagery experience "An Adventure in Space," a selection from the recording *On the Move with Steve and Greg* (Youngheart Records, 1983, Los Angeles, CA, order #YR0054). Have the children act out different parts of the story from countdown through travel in space and the return to Earth.

Planning a Space Mission

Assign children to a space shuttle crew of three to five, depending on the size of your classroom space shuttle.

Plan special missions to different planets, moons, or galaxies. Make a list of explorations and experiments the junior astronauts might want to conduct on their journey. What do they want to learn about the planet or galaxy? What will their work entail? How long will they be away? How much food should they take? Will they take turns sleeping and working at the controls? What problems might they encounter during the journey?

Design a space shuttle mission patch to wear in class. The patch should include the date, the astronauts' names, a mission number, and something special about the mission. For example, if children plan to explore Venus on their mission, their patch design should show something special about Venus.
materials: fabric or heavy paper patches, markers, thread, needles

Brainstorm a list of reasons why we might want a space treaty to govern activities of different groups that explore and use outer space. Share *The Outer Space Magna Charta* (see below) with children. Discuss implications of the treaty. Children should consider principles of the treaty as they plan their own journeys into space.

The Outer Space Magna Charta

"A Treaty of Principles Governing the Activities of States in the Exploration and Use of Outer Space, Including the Moon and Other Celestial Bodies."

1. International law and the charter of the United Nations shall apply to space activities.
2. Outer space and celestial bodies are the province of mankind and shall be used only for peaceful purposes and for the benefit of all mankind.
3. Nuclear weapons, weapons of mass destruction, military bases, and military maneuvers are banned from space.
4. Outer space shall be free for exploration, use, and scientific investigation.
5. There can be no claims of sovereignty or territory by nations over locations in space, by means of use or occupation or by any other means.
6. Jurisdiction over space objects launched from Earth shall be retained by the launching state.
7. Private interests are recognized as having freedom of actions in space, so long as a government or group of governments on Earth authorize and exercise continuing supervision over their activities. Signatory nations (78 at last count, including the United States and the Commonwealth of Independent States (formerly the Soviet Union) are therefore under a duty to oversee the activities of their citizens and commercial ventures in space.
8. Governments are liable for damage caused on Earth by their space objects.
9. Astronauts are envoys of mankind and are entitled to non-interference and all necessary assistance in distress.
10. The natural environments of celestial bodies should not be seriously disrupted, and Earth must not be contaminated by extra-terrestrial organisms.

Space Exploration

Experiment with writing treaties: a playground treaty, a lunchroom treaty, a bus ride treaty, and so on.

Pretend to sleep in space. When astronauts sleep in space, they strap themselves into sleeping bags so they won't float around the cabin. The strap covers their arms, so their arms don't float around while they sleep. They wear earmuffs and eyeshades to block noise and light. Experience sleeping with earmuffs and eye shades. Lie still for a few minutes. How does it feel?
materials: earmuffs, eye shades, blankets, strips of fabric

Provide children with many of the tools an astronaut might use in space: a headset, suction cups for moving around the cabin, suction cup attachments for shoes.

Construct telescopes and binoculars to use on the space shuttle.
materials: paper towel rolls, toilet paper rolls, paint, clear plastic wrap

Place a computer inside the classroom space shuttle or close enough to it so that it becomes an integral part of the children's dramatic play. Outstanding software is available for children to use as they learn about space. A special disk for recording data from their pretend space explorations and a word processing program for making space log entries during their journey are important.

Investigate space toys. Imagine what it would be like to play with a slinky, jacks, or marbles in the weight-lessness of space. Space toys that simulate action in space can be purchased for student exploration and observation. **Record** observations in space journals.
materials: slinky, space marbles, jacks

Investigating Food in Space

Experiment with the process of dehydration-rehydration. Taste samples of dehydrated apples, apricots, and banana chips. Compare the taste with fresh fruits. Conduct a blind taste test. Record results of the test. Create a graph to show how many children could identify the dried fruit.
materials: dried apples, apricots, bananas, fresh fruits, chart paper

Investigate rehydration with orange juice drink crystals. Carefully measure and record the amount of water used to rehydrate the drink. Discuss the process of dehydration and rehydration. Calculate how much water would be necessary to rehydrate orange drink for a team of six astronauts during 20 days in space.
materials: orange flavored drink crystals, water, plastic cups, spoons, measuring cups

Investigate other dehydrated foods: instant oatmeal, instant mashed potatoes, instant soup mix, dehydrated vegetables, mushrooms, chicken, noodles, etc. Rehydrate the foods. Weigh food before and after rehydrations. Compare weights. How does adding water to food change the weight of the food? Create a chart for recording comparisons. Compare dehydrated spices like parsley and onion with fresh samples.
materials: foods listed above, water, bowls, kitchen scales, chart paper

Dry an apple. Peel and cut an apple into six round slices. Arrange the apple slices on a wooden skewer at 2 cm intervals. Rest the ends of the skewer on the sides of a box and leave to dry. Check daily. Record observations. Weigh the apples before drying and after. Compare weights. What happened?
materials: apples, wooden skewers, scale, box

2 cm

Experience eating food as astronauts did on early space flights. Pour instant pudding into a self-sealing plastic sandwich bag. Cut off the tip of one corner of the bag and eat it by squeezing the pudding through the opening. Mashed potatoes and other vegetables can also be eaten in this manner.
materials: self-sealing plastic bags, pudding, other foods

Imagine the effect of Earth's gravity on water. Pour water from a cup, squeeze from a medicine dropper, or splash in a basin. What makes the water go in the same direction? List all the ways children use water throughout the day. What would happen if water floated around you instead of falling to the ground? How could you "catch" it? Astronauts use a vacuum in space to collect water.
materials: water, medicine dropper, cups, basin of water

Walk on a treadmill for exercise. Practice other forms of exercise that might be appropriate for space travellers: sit-ups, jumping jacks, stretching, jogging in place, and so on. Create a chart to record the number of sit-ups or jumping jacks or the amount of time spent jogging. Check pulse rate before and after exercise. Record and compare.
materials: stopwatch, chart paper

Investigating Communication in Space

Provide children with headsets and microphones to use as they pretend to communicate with Mission Control. Tell children that air is needed for sound to travel. How can astronauts communicate in space where there is no air? Discuss the use of radios and computers for communication in space.

Act out scenes between astronauts and Mission Control. Practice using concise terms for communication: "roger," "over," "go for launch," and so on.

Investigating Technology Spin-offs

Explore technology spin-offs from NASA research programs used in our daily lives. The voice-controlled wheelchair, a reading machine for the blind, insulating films, and nuclear magnetic resonance used for medical diagnostic purposes are just a few of the spin-offs described in the NASA brochure: "Space Program Spin-offs."

NOTES

Our Solar System and Beyond

Teacher Information

Space... The word "space" refers to the infinite emptiness that is all around Earth. The space between planets, stars, and galaxies is empty—void of matter. There are five classifications of space:

terrestrial space: starts at Earth's surface and extends out about 6441 km (4000 miles);

cislunar space: the space between Earth and the moon;

interplanetary space: begins about 80.5 billion km (50 billion miles) from Earth where the sun's gravity becomes stronger than Earth's;

interstellar space: space between the stars;

intergalactic space: never-ending space between the galaxies.

Distances in Space... Distances in space are measured in light-years. Light travels 300,000 kilometers per second (186,000 miles per second). A light-year is the distance that light travels at this speed in one year—approximately 9.6 trillion km (6 trillion miles). From our solar system, it is about 30,000 light-years to the center of the **Milky Way.**

The Universe... Universe means everything there is—the sun, the moon, the stars, and the planets. All of space and everything in space is part of the universe. Scientists debate whether there is an end to the universe.

Galaxies... Galaxies are huge groups of stars held together by mutual gravitation. Through a telescope, galaxies look like islands out in space. Galaxies can be spiral, elliptical, or irregular in shape. Within each galaxy, there are groups of stars known as constellations, other stars, and planets. Our galaxy is called the Milky Way. Scientists believe there are millions and millions of galaxies in the universe, but we don't know exactly how many.

Stars... Stars are huge, hot, shining balls of gas. All stars, like our sun, produce their own light by nuclear fusion reactions in their centers. If you look carefully at stars, you will notice they are different colors. Their colors give clues to their surface temperature. The hottest stars that you can see are blue-white with a surface temperature of over 10,000°C (19,000°F). The coolest stars appear red. Their surface temperature is about 2000°C (3500°F). Astronomers estimate that there are 200-billion-billion stars in the universe.

Constellations...A constellation is a group of stars that seem to form a pattern or dot-to-dot picture in the sky. Ancient people told stories about such constellations as Orion, the Hunter; Ursa Major, the Great Bear; and Pegasus, the Winged Horse. As Earth travels around the sun, our view of the stars in outer space keeps changing. The sun, moon, and planets seem to move among 12 constellations during the year.

Planets... Planets differ from stars in that they do not produce their own heat or light, but rather reflect the light from the sun. Planets also travel in orbits and rotate on their axes.

Asteroids... Asteroids are chunks of rocks of different sizes and shapes. Most are small chunks of rock less than one km (.62 mile) in diameter; the largest is less than 805 km (500 miles) wide. Asteroids are most abundant in the Asteroid Belt between Mars and Jupiter.

Comets... Comets are large balls of glowing gases, dust, and ice. They consist of a central mass surrounded by a misty envelope that may extend into a tail of glowing gases streaming out behind. They move around the sun like planets and asteroids, but in very erratic, long, cigar-shaped orbits. A comet can be over 1,600,000 km (one million miles) wide.

Meteorites and Meteors...A meteorite is a mass of stone and metal, often a part of a comet, that flies through space. A meteorite is called a meteor or shooting star when it enters Earth's atmosphere, creating great friction that results in heat and a luminous glow in the sky. Meteorites can be hazardous to space travel.

Our Solar System... Our solar system is composed of a large star, the sun, and nine known planets. A belt of asteroids also orbits the sun, in a path between Mars and Jupiter. Our solar system—the sun and all its planets—is traveling around the center of the galaxy at a speed of approximately 280 km (175 miles) a second.

The Seasons... Earth rotates on its axis—an imaginary line that goes through the North and South Poles. The axis is tilted 23.5 degrees, causing different parts of Earth to get varying amounts of sunlight as it revolves around the sun. When Earth's axis tilts most away from the sun on December 22, winter starts in the northern hemisphere and summer starts in the southern hemisphere. On June 21, Earth's axis tilts most toward the sun, causing the sun to shine directly on the northern hemisphere and causing summer to begin above the equator.

Our Solar System: Daylight Investigations

Investigating the Sun

Tell the story of Apollo, the sun god, who brings life-giving heat and light to Earth.

Introduce children to a portrait of the sun. Study the sun's surface features and a cross-section of its interior using the **T(A):** *A Portrait of the Sun.* (See instructions on inside front cover.) Investigate the process by which nuclear energy radiates from the core out to the convection zone where giant bubbles of hot gases rise and fall. Draw and label a portrait of the sun in space journals.

Explore the relationship between the sun and life on Earth. Provide children with many opportunities to understand that without the sun's energy, human life would not exist on Earth.

List all the ways we depend on the sun: winds, water cycle, seasons, and photosynthesis. Help children infer that sunlight is at the foundation of Earth's food chain. Record in space journals.

Explore ancient sun calendars like Stonehenge and Native American medicine wheels. Sketch designs and relationship to the summer solstice sunrise in space journals.

Experiment with the amount of heat or infrared radiation absorbed by different colors. Place ice cubes in three or four separate bowls. Place a different colored fabric—from very light to very dark— over each bowl and place them in direct sunlight or under an artificial, equally distributed light source. Check frequently. Record observations. Which ice cubes melted first? Last? Which colors absorbed the most heat? Which colors reflected the most heat? What colors might you wear to help you stay warm on a cold day or stay cool on a hot day? Why is the space shuttle white? Why are space suits white?
materials: different colors of fabric, bowls, ice cubes

Sun Facts

• It would require 333,000 Earths to equal the Sun's mass.

• The volume of the Sun is so great that a million Earths could fit inside.

• 109 Earths could fit side-by-side across the diameter of the Sun.

• Sun diameter– 1,392,000 km (8,644,320 mi.)

• Earth diameter– 12,756 km (7920 mi.)

Investigating Light and Shadows

Observe shadows cast by different objects—cars, trees, buildings, people, fences, and flagpoles. How is a shadow made? Can sunlight pass through your body? Can you identify an object by the shadow it casts?

Create shadow pictures. Trace around shadows created by different objects. Cut out and compare with objects.
materials: paper, scissors, sunshine or light source

Investigate visible light. Test different materials to determine what things light will pass through. Children can classify materials into two groups: opaque materials that keep out light and transparent or translucent materials that let light through. Collect different materials: aluminum foil, paper, cellophane, cardboard, wool, waxed paper, plastic, tissue paper, metal, wood, water, and so on. Record observations on **(R):** *Can Light Get Through?,* page 18.

Space Exploration

Investigate reflection of light with a mirror, a flashlight, and sunlight. Hold the mirror in the sunlight, shining its light onto a wall. What happens to the sunlight? Get a flashlight and shine it on a mirror. What happens to the light? Help children infer that light bounces off the mirror, much like a ball bounces off a wall. **Record** experiment, observations, and understanding in space journals.
materials: mirrors, a flashlight, sunlight

Demonstrate that the shape of a shadow changes because of Earth's rotation. Place a toothpick or match stick into a piece of modeling clay stuck on a world globe. Shine the light from an overhead projector or flashlight on the globe creating a shadow cast by the toothpick. Have children watch the shadow carefully as you rotate the globe. Watch the "sun" rise and set on the toothpick. What happens to the toothpick's shadow? Discuss observations.
materials: toothpick or match stick, clay, overhead projector or flashlight, world globe

Investigate the results of Earth's rotation by observing changes in children's shadows as the sun moves across the sky from east to west. Working in pairs, have children trace each other's shadow with chalk on a large sunny paved surface. Remember to take special care to trace around the shoes. After an hour, have children stand back on their shoe prints. Observe and trace the new shadow. What happened? Continue throughout the day. What would your shadow look like at sunset? Sunrise? **Record** in space journals.
materials: chalk, sunny day, large paved surface

More Shadow Activities

Observe the direction of shadows at a specific time. Do shadows point in the same direction at different times of day? Where is the sun? Help children discover that the shadow is on one side of an object and the sun is on the other side. Illustrate and describe observations in space journals.

Observe shadows moving as objects move. Play shadow tag. Can you touch your shadow? Can you step on your shadow with one foot? Two feet?

Explore shadows formed when a light source (the sun) is partially blocked by an object. Note that the direction and length of a shadow is determined by the position of the sun in the sky.

Observe what happens to your shadow when you stand in a shady area. What happens to the shadow? Stand with your shadow completely inside a larger shadow. Can you see the sun? **Record** in space journals. Can you see your shadow on a cloudy day? Why?

Design a shadow clock to investigate the shadows created by objects at different times of day. Can shadows help us tell the time? Make a shadow stick by placing a pencil in a lump of clay. Put the shadow stick in a place where the light from the sun can hit it for most of the school day. Place a large piece of construction paper under the shadow stick. Draw a half circle around the shadow stick using its length as a radius. Mark the position of the shadow each hour or half-hour at the point where it touches the circumference of the circle. Does it move? Also note the sun's position in the sky at each interval. When is the shadow longest? Shortest? *Important: Never look directly at the sun; it can severely damage your eyes.* **Record** investigation and observations in space journals.
materials: pencil, clay, paper, compass

Locate sundials in your community. Take a walk to one that is near your school. Does it resemble a clock?

Space Exploration

Investigating Night and Day

Explore the concepts of revolution and rotation. Explain to children that the planets in our solar system revolve around the sun. Place a large circle cut from yellow paper in the center of the room. Have children form a big circle around this "sun," join hands, and walk around the "sun" in a circle. Use the word "revolution" to describe this motion. Count complete revolutions. Time one complete revolution with a stopwatch.

Teach children the song, *Moving 'Round the Sun*, to sing as they revolve and rotate.

Moving 'Round the Sun

Sing to the tune of "This Is the Way We Wash Our Clothes."

This is the way we revolve 'round the sun, revolve 'round the sun, revolve 'round the sun.
This is the way we revolve 'round the sun. It takes us one full year. (Repeat.)

This is the way I rotate on my axis, rotate on my axis, rotate on my axis.
This is the way I rotate on my axis. It takes me one whole day. (Repeat.)

All of the planets move in this way, move in this way, move in this way.
All of the planets move in this way. All around the sun! (Repeat.)

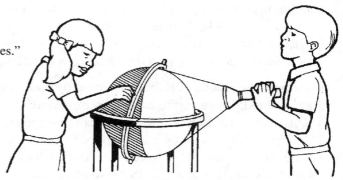

Demonstrate the rotation of Earth on its axis using a world globe. Learn that it takes Earth 24 hours, or one full day, to complete one rotation. Shine a light on the globe, and rotate the globe on its axis. An overhead projector is an excellent light source. Discuss observations. Children should infer that as Earth rotates, we experience night and day. Does the sun rise and set or does Earth rotate into and out of lightness and darkness? Encourage children to develop, test, and explain their own theories. Provide opportunities for children to repeat the demonstration. **Record** investigation and observations in space journals.
materials: globe, flashlight, overhead projector

Discuss the things we do during the day when the sun is shining and the things we do at night when our side of Earth is in darkness. List activities on a chart and draw illustrations.

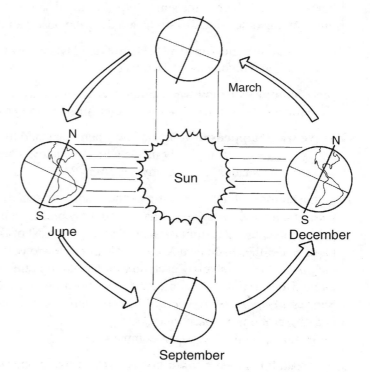

Investigating Seasons

Demonstrate the relationship between the tilt of Earth's axis and the seasons using a world globe and a light source. Call children's attention to the post on which the globe spins. It represents Earth's axis—an imaginary line that runs through the North and South Poles. Notice that the axis is tilted at 23.5 degrees on the globe stand, as it is in space. Have one child hold the light source while another child moves the globe around the "sun" demonstrating Earth's yearly period of revolution around the sun. Establish north in your classroom with a compass and instruct the child

Space Exploration

holding the globe to keep the northern tip of the axis pointing to the north at all times. Do children notice a difference in the amount of direct light on different parts of the globe at different points in its revolution? Are there any parts of the globe that get no light at specific points in Earth's orbital path? Provide children with opportunities to repeat the demonstration at other times to reinforce their understanding. Reinforce concepts with **T(B):** *The Seasons.* (See instructions on inside front cover.)
Record demonstration, observations, and understanding in space journals.
materials: world globe, overhead projector, compass, poster, or flashlight

Keep track of the length of days throughout the year using a newspaper or United States Naval Observatory Sunrise-Sunset Tables. Tables are published in *Nautical Almanac* issued yearly by the Nautical Almanac Office of the US Naval Observatory, available in public libraries. Encourage children to watch the sun rise and set. **Record** observations in space journals.

Investigating the Solar System with Telescopes

Provide children with magnifying glasses and encourage them to explore what happens to the image of an object when viewed through the lens. Does the image appear larger or smaller? Record images in space journals.
materials: a variety of magnifying glasses, objects for viewing

Observe the effect of distance on the apparent size of an object to help children begin to understand that although objects in the night sky look like tiny specks of light, they are many, many times larger than they appear. Locate an object in the distance, perhaps a tree, a building, lamp post, etc. Using the width of your hand, measure the height of the object. Walk toward the object, and measure the apparent change in size. Continue to walk toward the object measuring it at different points. What do you observe? Does the size of the object actually change? What might account for the different measurements?

Investigate concave and convex lenses. Observe images through the lens. Does the shape of the lens affect the image produced? How? **Record** observations in space journals.
materials: concave lens, convex lens, objects

Invite children to learn about the early astronomers who constructed telescopes for observing objects in the night sky in greater detail. Galileo constructed the first lens-refracting telescope and observed the moons of Jupiter, sun spots, and craters on the moon—things that are impossible to see with the naked eye. Investigate the work of the many scientists who contributed to our understanding of space. Write reports and share in a class book or in the class newspaper—*The Daily Sun.*

Collect pictures of different types of telescopes, and display in the classroom. Compare telescopes on Earth with the Hubble telescope.
materials: pictures of telescopes

NOTES

© Milliken Publishing Company
Space Exploration

Can Light Get Through?

Test different materials to find out which materials keep out light and which materials let light through.

    1. Collect a variety of materials.

    2. Hold the material to a light source to test for transparency.

    3. Classify by recording each material in the space below.

Transparent/Translucent	Opaque

Our Solar System: *Night Sky Investigations*

Investigating Constellations

Observe objects in the night sky on a clear evening. Record observations and share.

Share a map of constellations with children. Research the constellations visible in the night sky in your area during the unit of study. Create dot-to-dot illustrations of the constellations for children to use as a map to help them locate constellations in the night sky. Name the constellations and objects they represent.
materials: constellation map, paper, markers, or white tempera

Explain to children that if you observed the night sky for one year you would see some different stars each season. Study **T(C):** *Constellations in the Night Sky.* (See instructions on inside back cover.) Reinforce the concept by asking children to join hands in a circle and face out. Do they all see the same thing? Compare the child's view of the classroom to the view from Earth at different points in its revolution around the sun.
materials: T(C): *Constellations in the Night Sky*

Apply math skills: Create dot-to-dot constellation pictures using graph paper and coordinate points.
materials: graph paper, coordinate points for different constellations

Read ancient star myths to children. Invite children to retell the myths and dramatize the stories. Encourage children to create original star myths.

Explain that constellations are not flat, but rather three-dimensional. Compare to a mobile. Stars within any constellation may be separated from the rest by millions or billions of miles.

Make constellation mobiles with toothpicks and different sized styrofoam balls.
materials: toothpicks, styrofoam balls, string, hangers

Make constellation viewers. Cut circles out of black construction paper. Poke holes in the paper to form constellation patterns. Tape the paper over the end of a long cardboard tube. Label the tube with the name of the constellation and label which side is "up." Look through a tube into a light source and view the constellation. See reproducible **(R):** *Constellation Viewers,* page 20, for constellation patterns.
materials: paper towel rolls, black construction paper, pins, different sized nails, tape, light source

Build an astrodome in your classroom, a 3-D model of the stars in the night sky for children to assemble, available through Sunstone Publications. (See *Bibliography* for ordering information.)

Visit a local planetarium and view the night sky through the telescope. Look at the celestial objects with the naked eye after your visit and compare. See *National Geographic Picture Atlas of Our Universe* by Roy A. Gallant for a listing of planetariums and observatories in the United States and Canada. **Write** about the experience in space journals.

Space Exploration

Constellation Viewers

Name

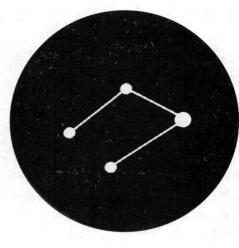

Orion (The Hunter)

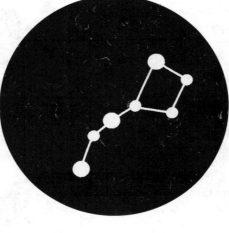

Big Dipper

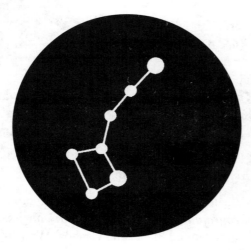

Little Dipper

Libra (The Scales)

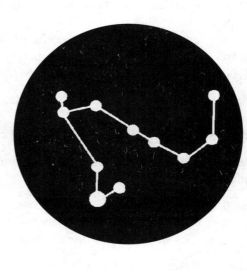

Draco (The Dragon)

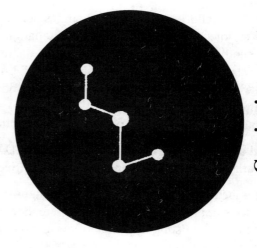

Cassiopia

Space Exploration

Invite an amateur astronomer to visit your class and talk about different types of telescopes and his or her favorite or most unusual observations of the night sky.

Investigating the Moon
Read *Papa, Please Get the Moon for Me,* by Eric Carle, aloud to the class.

Explore the different shapes of the moon during different phases. Introduce children to the terms: **full moon, new moon, first quarter, last quarter,** and **crescent.** Create the different shapes with modeling clay or salt/flour clay. Children can create moon necklaces by making a hole in the moon shape before it dries. Dry and paint. Thread a string through the hole and wear.
materials: modeling clay, flour, salt, colorful string, paint

New Moon	New Crescent	First Quarter	Full Moon	Third Quarter	Old Crescent	New Moon

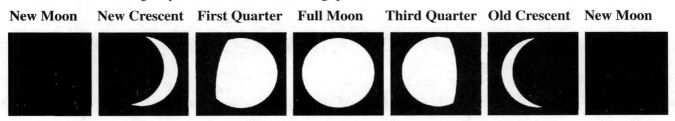

Demonstrate the phases of the moon with a flashlight, a world globe, and a white ball. The flashlight represents the sun, the globe represents Earth, and the ball represents the moon. Shine the flashlight or light source on Earth. One student holds the ball and moves around Earth to represent the moon's revolution. Discuss the meaning of **orbit.** Help children infer that the moon does not produce its own light, but rather reflects light from the sun and that the moon's position relative to Earth and sun accounts for the different phases of the moon. Note that each month the moon goes through this set of phases.
Record demonstration and understanding in space journals.
materials: flashlight, world globe, white ball

Observe the different phases of the moon over a one month period. Record observations on **(R):** *My Moon Calendar,* page 23. *This is a parent-child activity.*

Observe and record the movement of the moon in the night sky over a four-hour period. *This project is a parent-child activity.* Children must make their observations from a specific vantage point and choose an object in their neighborhood such as the chimney of their house or apartment building, a street lamp, a tall tree, etc., as a point of reference. Draw the vantage point and point of reference on a piece of paper. Observations should be made each hour over a four-hour period. Draw the position of the moon in relation to the reference point during each observation. Note the time of observation next to the moon drawing. Help children understand that the moon and Earth are moving in relation to each other. The motion we observe on one evening is largely the result of Earth's rotation on its axis. **Record** observations and understanding in space journals.
materials: paper, pencil, clock

Demonstrate that from Earth we see just one side of the moon. The moon spins on its axis (period of rotation) once every 27 days, 7 hours, and 43 minutes—the same amount of time it takes for the moon to orbit Earth (period of revolution). Because of this relationship, the same side of the moon always faces Earth. Demonstrate with a globe and a ball. Draw a line around the circumference of the ball, and label each side of the "moon."
Record demonstration and understanding in space journal.
materials: world globe, ball, tape

Space Exploration

Explore the surface of the moon with binoculars or a small telescope. What do the children see on the moon's surface? Look for the "man in the moon" or "green cheese." The best moon observations can be made when the moon is in its first quarter phase. The full moon is so bright that many surface features are lost in the glare. *This activity is a parent-child activity.*

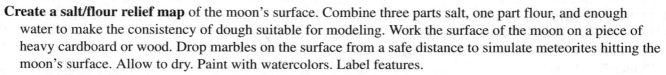

Study surface features of Earth's moon: Sea of Tranquility, Sea of Crises, Sea of Fertility, Sea of Nectar, Sea of Clouds, Ocean of Storms, Carpathian Mountains, Apennine Mountains, Sea of Rains, Bay of Rainbows, Jura Mountains, Northern Highlands, and Southern Highlands. Can you imagine why these names were given to geographical features on the moon? Are they named after similar geographical features on Earth? How is the surface of the moon similar to Earth's? Different? How might we account for these surface features on the moon? Volcanoes? Earthquakes? Meteorites? Research possibilities. **Record** in space journals.
materials: map of moon's surface or a moon globe

Create a salt/flour relief map of the moon's surface. Combine three parts salt, one part flour, and enough water to make the consistency of dough suitable for modeling. Work the surface of the moon on a piece of heavy cardboard or wood. Drop marbles on the surface from a safe distance to simulate meteorites hitting the moon's surface. Allow to dry. Paint with watercolors. Label features.
materials: salt, flour, water, cardboard, marbles, labels, toothpicks, watercolors

Study the Apollo missions to the moon. Watch the NASA video, *Apollo 11—The Eagle Has Landed.*

Study lunar rock samples available for loan from NASA to teachers to use with students. Samples are encased in clear plastic and can be viewed with a stereo microscope. The samples come with a complete package of classroom activities. To qualify for a lunar sample loan, teachers must attend a certification workshop. Contact your NASA education office for information.

Research information about the moons of other planets in the solar system. Are these moons like Earth's moon? How are they similar? Different? **Record** research in space journals.

Create a creature suited to living on the moons of different planets. What will the creature eat? How has it adapted to survive temperatures on the moon? How can it get around? Where could it live? Draw a picture of your creature.

NOTES

My Moon Calendar

Observe the moon every night for one month. Carefully draw and shade in the shape of the moon each night on your moon calendar. Record the date of each observation.

Month _____ Year _____

Sunday	Monday	Tuesday	Wednesday	Thursday	Friday	Saturday
○	○	○	○	○	○	○
○	○	○	○	○	○	○
○	○	○	○	○	○	○
○	○	○	○	○	○	○
○	○	○	○	○	○	○

Space Exploration

Our Solar System: *The Planets*

Investigating Planets in Our Solar System

Read stories about the Roman gods and goddesses whose names and stories give identity to our planets. Write about the myths in the space journals. Create original myths.

Explore information about planet Earth as a class. Learn about the liquid, solid, and gaseous parts of Earth. Research information about the inner core, outer core, mantle, Earth's crust, and the layers of the atmosphere: troposphere, stratosphere, mesosphere, thermosphere, ionosphere, and exosphere. What processes of change caused the surface features of planet Earth?

Create a bulletin board to organize and share information learned in research. Discuss why humans, animals, and plants can live on Earth. Use this study of Earth as a point of reference as you and the children explore the other planets in the solar system.

Explore the orbital paths of the planets around the sun using **T(D):** *Our Solar System*. (See instructions on inside back cover.)

Provide children with many opportunities to become planet experts. Each child or team of children can research and record information about their favorite planet. Record information on **(R):** *All About Planet _____*, page 26. Provide children with many audiences where they can share their expertise. Include **(R):** *All About Planet _____* in space journals.

Create a class planet chart for recording information. Display the chart in a prominent place in the classroom and encourage children to fill in information about different planets as they discover it in their research. Review the chart frequently to discuss new information. Discuss similarities and differences between planets.

Create a class planet mobile, showing planets in proportional size. Choose the size ratio that will work best given your classroom environment. One possibility is: diameter of 1 cm=1000 km, making Jupiter 1.43 meters and Earth 13 cm. Some children can help with the calculations. Provide children with a variety of materials, so they can create a model of the planet that represents the surface features symbolically. Hang the planets from the classroom ceiling in order from the sun. Hang labels from each planet featuring interesting facts about the planet.

materials: calculators, heavy string; other materials will vary according to children's designs

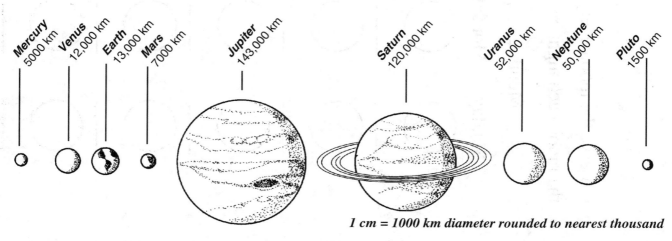

1 cm = 1000 km diameter rounded to nearest thousand

Space Exploration

Design a space colony. Which planet would provide the least hostile environment? Plan for oxygen supply, food supply, water, energy, gravity, housing, launching and landing areas, maximum population, waste management, recycling, and transportation from Earth as well as around the colony.
Record planning, design, and work in space journals.

Create a class planet trivia game. Provide children with blank note cards for recording interesting facts about the planets. Children will enjoy inventing games to play with the cards: a board game where each correct answer allows you to move ahead a specific number of spaces, a game of baseball where you get a "base hit" with each correct answer; a game where the goal is to get out of the solar system by answering at least one question about each planet. The possibilities are limitless and provide children with a fun way to increase knowledge about the planets.
materials: note cards, resource books, other materials children request as they create the games

NOTES

All About Planet _____

Planet goddess or god

_____

Planet symbol

Distance from the Sun _____

Period of Revolution _____

Period of Rotation _____

Diameter_____

Moons? _____ How many? _____

Average Temperature _____

Other interesting facts from your research:

_____

_____

_____

_____

_____

_____

Solar System Chart

Planet and Symbol	Average Distance from Sun	Rotation/ Revolution (in Earth Hours)	Diameter	Number of Moons	Composition	Atmosphere
Mercury	58 million km (36 million mi.)	59 days / 88 days	4880 km (3030 mi.)	0	Nickel-iron core about the size of the Earth's moon. Surface covered with craters; cliffs called rupes and scarps; and plains.	Almost no atmosphere. Small traces of helium, oxygen, and hydrogen.
Venus	108 million km (67 million mi.)	243 days / 225 days	12,100 km (7520 mi.)	0	Nickel-iron core in a rocky mantle and crust. Surface features mountains, rolling hills, volcanic craters, and plains.	Dense carbon dioxide atmosphere. Three layers of thick sulfuric acid clouds.
Earth	150 million km (93 million mi.)	24 hours / 365 days	12,756 km (7920 mi.)	1	Inner core: solid nickel-iron, outer core: liquid nickel-iron. Silicate mantle and crust. Three-quarters of surface covered with water.	Mainly nitrogen and oxygen, also contains small amounts of water vapor and other gases.
Mars	228 million km (142 million mi.)	24.5 hours / 687 days	6794 km (4220 mi.)	2	Iron sulfide core. Silicate mantle and crust. Reddish rocks, canyons, enormous volcanic mountains. Dust storms and polar caps.	Thin carbon dioxide atmosphere. Traces of other gases.
Jupiter	779 million km (484 million mi.)	10 hours / 12 years	142,984 km (88,750 mi.)	16	Small molten rocky core surrounded by thick layers of hot liquid hydrogen. Gaseous surface.	Several thick layers of brightly colored hydrogen clouds. Traces of helium, methane, and ammonia.
Saturn	1430 million km (890 million mi.)	10 hours / 29 years	120,536 km (74,560 mi.)	22	Small molten rocky core surrounded by water, ammonia, liquid metallic hydrogen, and liquid hydrogen. Gaseous surface.	1000 km deep atmosphere mainly composed of hydrogen with traces of other gases: ammonia, helium, and methane.
Uranus	2870 million km (1780 million mi.)	13–24 hours / 84 years	51,100 km (31,570 mi.)	15	Core of molten rock and ice surrounded by water and ammonia, hydrogen, and helium. Gaseous surface.	Murky air composed of mainly hydrogen with a little helium. Methane in upper atmosphere gives Uranus its green color. 8000 km deep.
Neptune	4500 million km (2790 million mi.)	18 hours / 165 years	49,200 km (30,200 mi.)	2	Core of molten rock and ice surrounded by water and ammonia, hydrogen, and helium. Gaseous surface.	Hydrogen, methane, and ammonia clouds. 8000 km deep. Appears bluish in color.
Pluto	5900 million km (3660 million mi.)	6 days / 248 years	3200 km (1900 mi.)	1	? Appears to be a world of frozen gases, perhaps covered with methane ice.	Thin atmosphere—most likely methane.

*Numbers on this chart are close approximations. The number of moons and temperature ranges of the outer planets will change as astronomers continue to learn more about the solar system.

Space Exploration

Energy, Force, and Motion in Space

Teacher Information

Gravity... Gravitation is the force of attraction that acts between all objects because of their mass. Mass is the amount of material or matter in any object. The weight is a measure of the attraction between two objects due to their mass. All objects with mass exert a force of gravity on each other. The strength of the force expressed as weight shows the relationship between the mass of the objects and the distance between them.

Gravity in our Solar System... The immense gravitational force of our nearest star, the sun, holds together the nine planets of our system. The planets travel through space at a speed that just balances the sun's gravitational pull, locking them into a perpetual circular orbit around the sun. This balance of forces—the outward or straight line thrust of any object and the gravitational pull between two objects is called **centripetal force**. The mass of the sun is so great—4.4 million, billion, billion, billion pounds—that its gravitational pull is enough to hold Pluto in orbit from a distance of 6 billion km (3.7 billion miles). Satellites and spacecraft orbit Earth in the same way.

Centripetal Force... Centripetal force is the force that keeps planets in orbit around the sun. The combination of the planets' forward motion and the inward pull of the sun's gravity (centripetal force) forces the planets to follow an orbit around the sun instead of flying straight out into space. This force also keeps moons in orbit around the different planets.

Escape Velocity... The speed at which a spacecraft must travel in order to overcome the pull of Earth's gravity.

How do airplanes fly? Wings are lifted by air flowing above and below them as they cut through the air. Air that pushes over the top of a wing speeds up and stretches out, causing the pressure above the wing to decrease. Air beneath the wing slows, causing the pressure to increase. The wing is pushed up from below, creating lift.

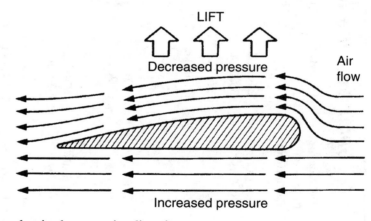

Action and Reaction... Newton's third law of motion states that for every action there is an equal and opposite reaction. This law applies to how a rocket works. Burning gases rush out from the rear of a rocket engine, propelling the rocket in the opposite direction.

Energy from the Sun—Radiation, Light, and Heat... We learn about the sun, stars, and galaxies by studying the light and other energy they radiate into space—the electromagnetic spectrum. Visible light is one form of radiation or energy emitted by stars and galaxies. Visible light can be broken up into a rainbow of colors called the visible spectrum. Each color of light has a different wavelength or energy; violet light has the most energy and the shortest wavelength, red light has the least energy and the longest wavelength. Visible light is only a very small portion of the energy emitted by stars and galaxies. Other forms of radiation emitted from stars and galaxies include gamma radiation, X-rays, ultraviolet, infrared, and radio waves. Scientists have developed telescopes that collect each type of radiation so they can learn more about our universe. Light waves travel in straight lines. When light hits an object, some waves are absorbed, others are reflected back out into space. The moon and the planets do not produce light and energy; they reflect light.

Energy, Force, and Motion: Investigations

Investigating Gravity

Experiment with the force of gravity by dropping different objects—pebbles, leaves, balls, feathers, paper, etc., from a high spot like a jungle gym. What happens? Do any of the objects go up? Why not? Tell children that we call this "falling force" gravity.
materials: objects for dropping

Demonstrate the effect of air resistance on falling objects with a piece of paper and a book. Hold both objects above the ground at arm's length with the larger surface parallel to the ground. Release both objects at the same time. What happens? Place the piece of paper on top of the book and drop the objects together. Does the paper fall at a different rate in both experiments? Discuss observations. **Record** experiment, observations, and understanding in space journals.
materials: paper, book

Organize a paper race. Experiment with dropping paper to the floor in as many different ways as possible— edge down, horizontally, crumpled. Record falling times. Can you make your paper fall very slowly? Discuss the reason for the different falling rates. Help children infer that the flat sheet of paper is slowed in its fall because it is pushed up by the air. The crumpled paper, because of its smaller surface area, hits less air and falls more quickly. Encourage children to find their own ways of explaining this concept.
Record experiment, observations, and understanding in space journals.
materials: paper in a variety of weights and sizes, stopwatch

Experience the physical effects of gravity. Ask the children to raise their arms to shoulder height and see how long they can keep them there. What happens? Why? What happens when you hang upside-down on the playground bars? **Record** experience in space journal.

Walk up and down a staircase. Repeat the process until the children are feeling a little tired. A long staircase is most dramatic. Is it more physically taxing to walk up the stairs or down? Why?

Note the force of gravity when building blocks tumble down, raindrops fall, snowflakes fall, and when children drop things.

Explore the meaning of the words *up* and *down.* Help children infer that up and down are totally relative to Earth's center of gravity and are not absolute directions like north, south, east, and west. Everyone on Earth thinks of up as toward the sky and down as toward the ground.

Investigating Aerodynamics—Flight within the Atmosphere

Brainstorm a list of things that fly: birds, insects, flying fish, plants, balloons, airplanes, gliders, helicopters, etc. Research the principles that allow each to fly. Do these things need the air in the atmosphere in order to fly? Could they fly in a vacuum?

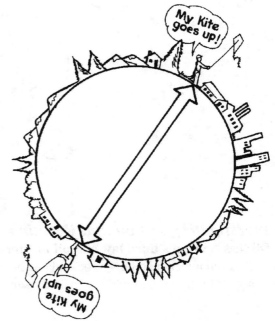

Space Exploration

Investigate paper gliders. Children can fly them around the school playground or in the school gymnasium to observe how they move through air. **Record** designs and experiments with flight in space journals.
materials: a variety of paper—different weights, sizes, colors; a book of glider designs

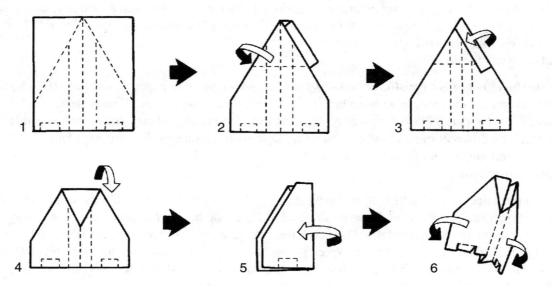

Change different variables to affect the flight of a glider. Measure and compare the distances traveled and flight duration of different glider designs. **Record** variables tested in space journals.

Compare a glider flight with the motion of a similar-sized piece of paper that is crumpled up and thrown. Why does the glider stay up longer? **Record** observations and understanding.

Demonstrate the concept of lift with a piece of paper about an inch wide and six inches long. Bend it slightly to resemble a wing. Blow over the top of the "wing." What happens? Why? **Record** experiment, observations, and understanding of the concept of lift in space journals.

Investigating Action and Reaction

Discuss Newton's third law of motion—for every action there is an equal and opposite reaction—to help children understand how a rocket launches a spaceship. Help children infer that rockets headed for space do not really fly in the same way a bird or an airplane flies, but rather use tremendous force to escape Earth's

Space Exploration

gravity. Note that the blasting flames and gases propelled out the end of the rocket are the "action," and the rocket's forward motion is the opposite "reaction."

Watch videos showing countdowns and blast-offs from Cape Canaveral in Florida available from NASA. Discuss observations.

Decorate a small paper bag to resemble a spacecraft. Attach it to a balloon at least 20 cm (8 inches) long. Blow up the balloon. Release. What happens? Help children infer that when the air escapes from the balloon, it pushes against air molecules outside of the balloon, which respond with an equal and opposite force, pushing the spacecraft forward. Attach the spacecraft to a larger rocket (balloon). What happens?

Investigate action and reaction with balloon rockets. Build a flight path for your paper bag-balloon rocket. Place a straw on a long string. Connect the ends of the string to objects in two different rooms. Tape the paper bag rocket to the straw. Inflate a 20 cm (8-inch) balloon, and place it inside the bag. Let go. Measure how far the rocket travels. Experiment with different sizes of balloons. What happens?

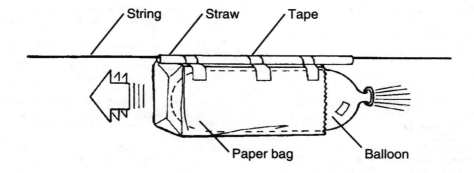

Bibliography

Astronauts. Carol Greene (Children's Book Press, 1984).

Astronomy. Dennis B. Fradin (Children's Book Press, 1983).

Astronomy Encyclopedia. Frances M. Clapham and Ron Taylor (Rand McNally & Co., 1984).

Astronomy Handbook. James Muirden (Arco Publishing, 1982).

Astronomy: Planets, Stars and the Cosmos. H. Couper, and N. Henbest (Aladdin Books Ltd., 1983).

Astronomy Today: Planets, Stars, Space Exploration. Dinah L. Moche (Random House, 1982).

Atlas of the World. B.R. Ernst, TJE De Vries (Nelson, 1961).

A Spotter's Guide to the Night Sky. Nigel Henbest (Mayflower Books, 1979).

Book of Greek Myths. D'Aulaire (Doubleday, 1962).

Constellations: How They Came to Be. Roy Gallant (Macmillan, 1979).

Comets. Franklyn M. Branley (Harper & Row, 1874).

Comets, Asteroids, and Meteors. Dennis B. Fradin (Children's Book Press, 1984).

The Cosmic Mind Boggling Book. Neil McAleer (Warner Books, 1982).

The Creation of Atoms and Stars. David Fisher (Holt, Rinehart and Winston, 1979).

Earth—Our Planet in Space. Seymour Simon (Macmillan Child Group, 1984).

Eclipse: Darkness in Daytime. Franklyn M. Branley (Thomas Y. Crowell Co., 1973).

Exploring the Night Sky. Terence Dickinson (Camden House Publishing, 1987).

Exploring the Night Sky by Day. Terence Dickinson (Camden House Publishing, 1988).

Fires of Life. (Smithsonian Exposition Books, 1981).

Galaxies. Seymour Simon (Mulberry, Morrow, 1991).

How to Be a Space Scientist in Your Own Home. Seymour Simon (Lippincott, 1982).

How We Learned the Earth is Round. Patricia Lauber (Harper C. Child Books, 1990).

Journey to the Planets. Patricia Lauber (Crown Publishers, Inc. 2nd ed., 1990).

Jupiter. Seymour Simon. (William Morrow and Co., 1985).

Let's Visit a Space Camp. Edith Alston, (Troll Assocs., 1990).

The Long View into Space. Seymour Simon (Crown, 1987).

Looking to the Night Sky: An Introduction to Star Watching. Seymour Simon (Puffin Books, 1977).

Mars. Seymour Simon (Morrow, 1987).

Micromodels: Make Your Own Stephenson's Rocket. Myles Mandell (Putnam Pub. Group, 1983).

The Moon. Seymour Simon (Macmillan Child Group, 1984).

My First Book of Space. Rosanna Hansen and Robert A. Bell (Simon & Schuster, 1985).

Mysteries of Outer Space. Franklyn M. Branley (Lodestar Books, 1985).

National Geographic Picture Atlas of Our Universe. Roy A. Gallant (National Geographic Society, 1980).

The New Moon. Herbert S. Zim (William Morrow and Co., 1980).

Neptune. Seymour Simon (Morrow, 1991).

The Night Sky Book. Jamie Jobb (Little, Brown and Co., 1977).

Of Quarks, Quasars, and Other Quirks: Quizzical Poems for the Supersonic Age. Sara and John E. Brewton and John Brewton Blackburn (Thomas Y. Crowell, 1977).

Once Upon a Starry Night. Von Del Chamberlain (Hansen Planetarium, 1984).

101 Questions and Answers About the Universe. Roy A. Gallant (Macmillan Publishing Co., 1984).

The Paper Airplane Book. Seymour Simon (Puffin Books, 1976).

The Planets: Exploring the Solar System. Roy A. Gallant (Four Winds Press, 1982).

The Planets in our Solar System. Franklyn M. Branley (Thomas Y. Crowell, 1981).

The Practical Astronomer. Colin A. Ronan (Macmillan Publishing Co., 1981).

Saturn. Seymour Simon (William Morrow and Co., 1985).

Seeing Earth from Space. Patricia Lauber (Orchard Books, 1990).

The Sky Is Full of Stars. Franklyn M. Branley. (Thomas Y. Crowell, 1981).

Sky Watchers of Ages Past. Malcolm E. Weiss (Houghton Mifflin Co., 1982).

Solar System (from the Planet Earth Series). Kendrick Frazier (Time-Life Books, 1985).

Space Scientists Projects for Young Scientists. David McKay (Watts, 1989).

Space, Time and Infinity. James S. Trefil (Smithsonian Books, 1985).

Space Voyager. Wendy Boase (Walker Books, 1984).

Space Words—A Dictionary. Seymour Simon (Harper C. Children's Books, 1991).

The Star Book. Robert Burnham (Cambridge University Press, 1984).

Stars. Seymour Simon (Morrow Jr. Books, 1986).

The Stars: A New Way to See Them. H.A. Rey (Houghton Mifflin, 1976).

The Stargazers' Bible. W. S. Kals (Doubleday, 1980).

Star Signs. Leonard Everett Fisher (Holiday House, 1983).

The Story of the Challenger Disaster. Zachary Kent (Children's Book Press, 1986).

The Sun. Seymour Simon (Morrow Jr. Books, 1986).

Time After Time. Melvin Berger (Coward, McCann & Geoghegan, Inc., 1975).

Uranus. Seymour Simon (Morrow, 1990).

The Young Astronomer. Christopher Maynard (Usborne Publishing, 1977).

Whitney's Star Finder. Charles A. Whitney (Alfred A. Knopf, 1981).

The World's Space Program. Isaac Asimov (Gareth Stevens Inc., 1990).

Teacher Resources

Astronomy on a Shoestring. National Science Teachers Association, 1742 Connecticut Ave. NW, Washingnton, DC 20009.

Astronomical Education Materials Resource Guide. Astronomy Education Materials Network, Dept. of Curriculum and Instruction, West Virginia University, Morgantown, WV 26506.

The Astronomical Society of the Pacific. 1290 24th Ave., San Francisco, CA 94122.

Astro-Dome. Sunstone Publications, PO Box 788, Cooperstown, NY 13326.

Hansen Planetarium Publications. 15 S. State St., Salt Lake City, UT 84111.

NASA. For publications list and location of resource centers write: Educational Publications and Special Services Branch, LEP, NASA, Washington, DC 20546.

National Geographic. Education Services, Dept. 38, Washington, DC 20036.

Star Challenger—Guides & Games for Star Gazers. Discovery Corner, Lawrence Hall of Science, University of California, Berkeley, CA 94720.

The Young Astronaut Council. PO Box 65932, Washington, DC 20036.